SWU-600-001

THE EXERCISE OF ARMES

✤

BY JACOB DE GHEYN II

Series curated by
Luca Stefano Cristini

SOLDIERSHOP PUBLISHING

AUTHOR

Jacob de Gheyn II (1565 - 1629). Dutch renommated artist and engraver was born in Antwerp and died in The Hague (NL). He was a pupil of Hendrick Goltzius in Haarlem. He is one of the circle of engravers of the Mannerist school like Matham, Saenredam and Goltzius. De Gheyn made also interesting prints on military items. His series of military exercises, subjets of our book is widely known.

PUBLISHING'S NOTE

NOTE ABOUT BOOK PRINTING BEFORE 1925

LICENSES COMMONS

ACKNOWLEDGEMENTS

A Special Thanks to Rijkmuseum and other institutions for their kindly permission or policy to use some images of his archives, collections or books used in our book.

Title: **THE EXERCISES OF ARMES - by Jacob de Gheyn II**
Series edit & curated by Luca S. Cristini. First edition by Soldiershop. September 2016
Cover & Art Design: Luca S. Cristini. Plates re-colorations by Anna Cristini.
ISBN code: 978-88-93271240
Published by Soldiershop publishing, via Padre Davide, 7 - 24050 Zanica (BG) ITALY. www.soldiershop.com

SOLDIERSHOP
PUBLISHING
BOOK on DEMAND

THE EXERCISE
OF ARMES

*

BY JACOB DE GHEYN II

HISTORICAL ILLUSTRATIONS OF THE DRILL
IN THE USE OF MUSKET, CALIVER AND PIKE

The celebrated Dutch artist Jacob de Gheyn put his considerable talents to work in this 1607/1608 masterpiece, one of the earliest and most famous manuals of arms ever published.

Eited in several language, it appear in Holland with the first title of *Wapenhandelinghe van Roers, Musquetten ende Spiessen.* Named: *Maniement d'armes, d'arquebuses, mousquet et piques* in French language, and *Exercise Of Arms* in England.

Its 117 handsome copper engravings, with their captions, magnificently portray the step-by-step sequence for training foot soldiers in the handling of the standard weaponry of the XVII century warfare: muskets, matchlock or calivers, and pikes.

This work of De Gheyn was realized in 3 great chapitres: 42 plates for the caliver (small arquebuses), 43 for the musket, and last 33 plates for yhe use of pike in infantry tactics.

This work of Jacob De Gheyn result of great importance also for the future great Rembrandt for his master art : The night watch.

A priceless resource for the organization and training of troops, de Gheyn's book created an overnight sensation throughout Europe and was quickly translated into several languages.

An incredible number of imitations editions have since appeared, but *The Exercise of Armes* remains the classic. Above and beyond its intrinsic historic interest, the volume presents a meticulously accurate portrait of uniforms and weapons of the era of Netherlands and Europe, in addition to the aesthetic appeal of its remarkable engravings of this great artist!

◀ *Portrait of Maurice of Nassau, Prince of Orange paint by Michiel van Mierevelt (1567-1641)*

CONTENTS

*

*

JACOB DE GHEYN II AND THE DUTCH MILITARY ART IN 17th CENTURY

De Gheyn was born in Antwerp in 1565 and received his first training directly from his father, Jacob de Gheyn I, a glass painter, engraver, and draftsman. In 1585, he moved to Haarlem, where he studied under Hendrik Goltzius for the next five years, and of which become the most important pupil. He is one of the great artist of the circle of engravers of the Mannerist school like Matham, Saenredam and Goltzius.

In this years he absorbed Goltzius's sinuous linear technique, which appeared in de Gheyn's early engravings. Mannerism is the style between the late renaissance and early baroque. Typical is a certain exaggerated tortion of the figures, often with thick fingers and cheeks, and small mouth with filled lips. The scene is mostly placed in an imaginary searched landscape with a prominent tree with twisted branches under a tormented sky. This is the cultivation of the art feeling of late 16th century.

After Harlem, Jacob moved again, to Leiden, in the middle of the 1590s. His work attracted the attention of wealthy sponsors, and his first commission was for an engraving of the Siege of Geertruidenberg from Maurice of Nassau, Prince of Orange.

This event from March 27 to June 24, 1593, had been more of a demonstration of power by Prince Maurits, than an actual war, and had even attracted tourists. As a publicity stunt, the siege and its subsequent engraving were successful in propagating an image of Prince Maurits as an able general.

Around 1600, de Gheyn abandoned engraving, and focused on painting and etching. Moving to The Hague in 1605, he was employed often by Dutch royalty, designing a garden in the Buitenhof for Prince Maurice of Orange which featured the two first grottoes in the Netherlands. After Prince Maurice's death in 1625, de Gheyn worked for his brother, Prince Frederick Henry. De Gheyn painted some of the earliest female nudes, vanitas, and floral still lifes in Dutch art.

He is credited with creating over 1,500 drawings, including landscapes and natural history illustrations. While in Amsterdam, in the years 1607/1608, he made 117 designs for engraved illustrations in a military training manual to aid the Dutch fight for independence from Spain. This military manual is: *The Exercise of Armes* presented in this book.

De Gheyn married Eva Stalpaert van der Wiele of Mechelen in 1595. His son, Jacob de Gheyn III, was born in 1596, and grew to become an engraver in his own right, as well as the subject of a portrait by Rembrandt. De Gheyn died in The Hague in 1629.

THE WORK THAT INFLUENCED REMBRANDT...

The first editon of 1608 of this famous military manual had a finest frontispices and 118 copper engravings plates.

The wonderful enravings of De Gheyn has influenced Rembrandt for his famous painting "The Night Watch". *"Rembrandt borrowed poses from the then well-known 'Exercise of Arms' with engravings by Jacob de Gheyn of 1608. He presented the three most important exercises the civic guards engaged in in logical order from left to right on the painting's central plane: 1. loading. 2. firing. 3. blowing residual gunpowder away from the firing pan."*

The illustration in the book is divised in three parts : 42 plates concerning firelocks, 43 plates of muskets and 33 plates of pikemen.

The same De Gheyn manual had influenced another oldest work on the same subject realized in 1618 by Adam van Breen (Amsterdam, 1585 - 1642), an other Dutch Golden Age painter: *Le maniement d'armes de Nassau, avec rondelles, piques, espees et targes.*

▲ *The night watch by Rembrandt - Rijksmuseum, Amsterdam*

IACOBUS DE GEYN, ANTVERP.
PICT. ET SCULPT.

▲ *Jacob de Gheyn in a contemporary engraving.*

This military manual of De Gheyn provides instructions for handling pikes and muskets, together with standardized commands for drill masters. Its most important feature, however, is its wonderful engraved illustrations. These provide step-by-step guidance for using weaponry, as well as depicting contemporary military dress. The elegance of the first editions and the use of large folio format suggests that it was not necessarily affordable for soldiers, but was aimed at high officers in command of militias.

Published in multiple editions and several European languages, it testified to the worldwide reputation of the Netherlands as *"the nurserie of soulderie"* (in the words of Henry Hexham, author of a similar manual).

Dutch military reform had been fostered during the revolts against Spain, during the 80 years war under Maurice, Prince of Orange. He promoted classical drill discipline, which was to influence military training expecially during the Thirty Years' War and the English Civil War.

PIKE AND SHOT

Pike and shot is a historical infantry combat formation, that is generally considered evolved in the period from the Italian Wars to the evolution of the bayonet in the late seventeenth century. The infantry formations of the period were a mix of pike and early firearms ("shot"), either arquebusiers or musketeers. Later called also caliver and matchlock.

▲ *Prince Maurice of Orange during the Battle of Nieuwpoort (1600) by Henri Ambrosius Pacx*

THE DUTCH REFORM

Foremost amongst the enemies of the Spanish Habsburg empire in the late 16th century were the Seven Provinces of the Netherlands (often retroactively known as the "Dutch"), who fought a long war of independence from Spanish control starting in 1566. After soldiering on for years with a polyglot army of foreign-supplied troops and mercenaries, the Dutch took steps to reform their armies starting in 1590 under their captain-general, Maurice of Nassau, who had read ancient military treatises extensively.

In addition to standardizing drill, weapon caliber, pike length, and so on, Maurice turned to his readings in classical military doctrine to establish smaller, more flexible combat formations than the ponderous regiments and tercios which then presided over open battle.

Each Dutch battalion was to be 550 men strong, similar to the size of the ancient Roman legionary 480-man cohort described by Vegetius. Although inspired by the Romans, Maurice's soldiers carried the weapons of their day, 250 were pikemen and the remaining 300 were arquebusiers and musketeers, 60 of the shot serving as a skirmish screen in front of the battalion, the rest forming up in two equal bodies, one on either side of the pikemen. Two or more of these battalions were to form the regiment, which was thus theoretically 1.100 men or stronger, but unlike the tercio, the regiment had the battalions as fully functional sub-units, each of mixed pike and shot which could, and generally did, operate independently, or could support each other closely.

These battalions were fielded much less deep than the infantry squares of the Spanish, the pikemen being generally described as five to ten ranks deep, the shot eight to twelve ranks. In this way, fewer musketeers were left inactive in the rear of the formation, as was the case with tercios which deployed in a bastioned-square.

Maurice called for a deployment of his battalions in three offset lines, each line giving the one

▲ *From the "Les miseres et les malheurs de la guerre," of Jacques Callot, 1633*

in front of it close support by means of a checkerboard formation, another similarity to Roman military systems, in this case the Legion's Quincunx deployment.

In the end, Maurice's armies depended primarily on defensive siege warfare to wear down the Spanish attempting to wrest control of the heavily fortified towns of the Seven Provinces, rather than risking the loss of all through open battle. On the rare occasion that open battle occurred, this reformed army, as many reformed armies have done in the past, behaved variably, running cravenly from the Spanish tercios one day, fighting those same tercios only a few days later, at the Battle of Nieuwpoort, and crushing them. Maurice's reforms are more famous for the effect they had on others—taken up and perfected, and would be put to the test on the battlefields of the seventeenth century.

Jan tjel ex,   Manderen Jnuen   N G

Marte novo exultat cristasq invertie iactat   Sulck Crijghsman ionck hem vroolijck op den tocht gheeft
Mites, inxperta qui capit arma manu   Want Crygh is soet voor die hem noyt besocht heeft

▲ The soldier's life in XVII century by Gillis Van Breen

▲ Portrait of Frederick Henry, Prince of Orange, of Michiel van Mierevelt (1567–1641)

MANIEMENT D'ARMES
D'ARQVEBUSES, MOUSQVETZ,
ET PIQVES.

EN CONFORMITE DE L'ORDRE DE
Monseigneur le Prince Maurice, Prince
d'Orange, Comte de Nassau &c. Gouver-
neur et Capitain General de Geldres,
Hollande, Zeelande, Utrecht,
Overyssel &c.

REPRESENTE PAR FIGURES, PAR
Iaques De Gheyn.
Ensemble les enseignemes par escrit
A l'utilite de tous amateurs des armes et aussi
pour tous Capitaines & commandeurs, pour par ceux
pouvoir plus facilement enseigner a leurs soldatz
inexperimentez, l'entier et parfait maniement d'icelles armes.

Imprime a Amsterdam chez Robert de Baudous, avec
Privilege de l'Empereur, du Roy de France, &
des Nobles & Puissans Seigneurs Mes-
seigneurs les Estatz generaulx
des Provinces Vnies.

1608

Illustrissimo Principi, maximo duci, Mauritio
Nassavio Pr Arauf. copiarum Federati Belgii
imperatori, disciplinæ militaris parenti has scul-
tæ ab ipso armaturæ effigies salutaris curæ
monumentum, dicabat Iacobus Geinius.
On les vend aufsi a Amsterdam chez Henry Laurens.

THE
COLOUR
PLATES

1

THE
CALIVER DRILL
SECTION

1- Shoulder your caliver

2 - Unshoulder your caliver

3 - with the right hand hold the lit fuse

4 - The caliver in your left hand

5 - Take the caliver in your right hand

6 - Blow on fire of your caliver

7 - Cock your caliver

8 -   *check your weapon*

9 - Blow your caliver and open your part

10 - Present your weapon

11 - Just presentation give fire!

12 - Balance your caliver in left hand and take separet the muzzle

13 - Uncock your caliver

14 - *Again the caliver in your left hand*

15 - *Blow out the pan and take your flask with right hand*

16 - Insert the powder

17 - Close your caliver pan

18 - Shake off the remained powder

19- *Blow up any loose powder*

20 - Take the caliver in your hands

21 - Balance the caliver with you left arms and take the flask with the right hand

22- Open a new charhe for your caliver

23 - Charge your weapon

24 - Draw out your scouring stick

25 - *Take your scouring stick to prepare load a new ball*

26 - Complete your charge

27 - Withdraw your scoring stick

28 - Shorten your scouring stick

29 - Place in side your scouring stick

30 - Shoulder your caliver (first part)

31 - Shoulder your caliver (second part)

32 - Shoulder your caliver (third part)

33 - The caliver correctly on your shoulder

34 - Unshoulder your weapon

35- Control the weapon

36- Hold your caliver in correct way

37 -Take the caliver only with the left hand

38 - Take your weapon in your right hand

39 - Blow your match

40 - *Cock your match*

41 - Try your match

42 - *Guard right and stand ready to fire*

2

THE
MUSKET DRILL
SECTION

IGheijn in

1 - Muskeeter in march with the weapon shouldered

2 - Shoulder your musket and carry the rest

3 - Unshoulder your musket

4 - hold the musket with right hand and the sink with the left

5 - Hold the musket in your left hand

6 - Take you weapon in the right hand

7 - Blow your match

8 - *Cock your match in the musket*

9 - Try your match

10 - Hold the musket and blow your match

11 - *Prepare your musket to the fire*

12 - Give fire to your weapon

13 - Take your musket an rest

14 - Uncock your match

15 - *Place correctly your match*

16 - *Blow out the powder*

17 - Prime your pan in the musket

18 - Shut your pan

19 - Cast off the excedent powder

20 - Blow out the loose powder

21 - Present out your musket

22 - Trail your rest

23 - Open a new charge

24 - *Charge your musket*

25 - Draw out your musket stick

26 - Shorten the stick and load a ball

27 - Complete your charge

28 - Withdraw your scouring stick

29 - Shorten the musket stick

30 - Place the stick in hi side

31 - Take the musket in your left hand

32 - With the musket in your right hand and the stick in the left

33 - Shoulder your musket

34 - *March and carry your weapon*

35 - Unshoulder your musket

36 - Place the musket in the rest

37 - Hold your musket on the rest

38 - Take your weapon well balance in the rest

39 - Take the match in your right hand

40 - *Blow your match*

41 - Cock your match

42 - Try your match

43 - Guard your pan and be ready to fire

3

THE
PIKE DRILL
SECTION

Ihrheijn in.

1 - The pikemen commander

2 - Advance your pike 1st

3 - Advance your pike 2nd

4 - with the pike right in your hand

5 - Order your pike 1st

6 - Order your pike 2nd

7 - The pike ordered

8 - Shoulder your pike 1st

9 - Shoulder your pike 2nd

10 - The pike shouldered at level

11 - The pike shouldered

12 - Port your pike 1st

13 - Port your pike 2nd

14 - Charge the pike

15 - Order your pike 1st

16 - Order your pike 2nd

17 - With the pike ordered

18- *The pike advanced*

19 - *The pike charged*

20 - Cheek your pike

21 - Trail your pike

22 - Charge the pike with two hands

23 - Continue charge the pike with two hands

24 - With the pike charged

25 - Charge your pike versus horse

26 - Shoulder the pike with the spike to the rear

27 - Prepare the pike 1st

28 - Prepare the pike 2nd

29 - Prepare the pike 3rd. The pike is charged

30 - Face about and shoulder the pike 1st

31 - Face about and shoulder the pike 2nd

32 - *Face about and shoulder the pike and, now the pike is shouldered*

SOLDIERS, WEAPONS & UNIFORMS ALREADY PUBLISHED
(SOME TITLES)

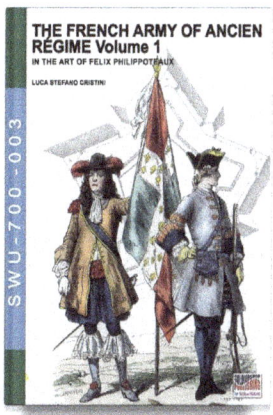
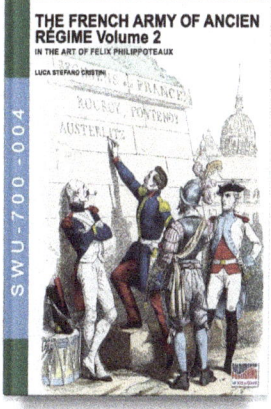
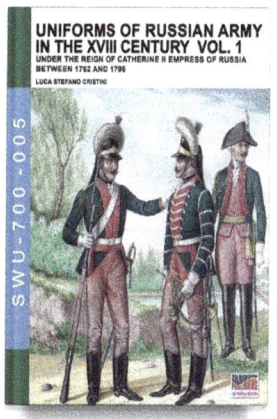

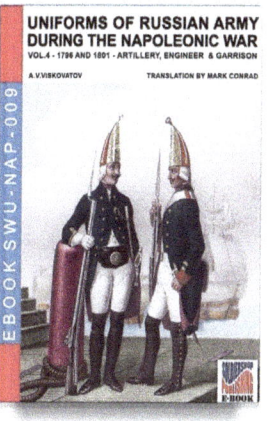

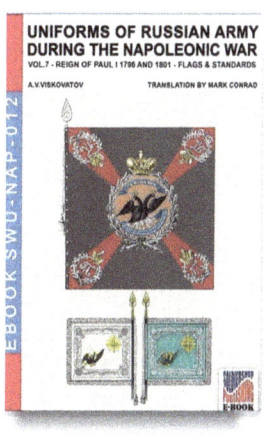

www.ingramcontent.com/pod-product-compliance
Lightning Source LLC
Chambersburg PA
CBHW041141120626
46547CB00020B/3074